FLORA OF TROPICAL EAST AFRICA

SALVADORACEAE

B. Verdcourt

Trees or shrubs, unarmed or with axillary spines. Leaves mostly coriaceous, opposite, simple, entire; stipules minute or absent. Flowers regular, hermaphrodite or dioecious, in dense axillary or terminal fascicles or panicles. Calyx-lobes 2–4(–5*). Petals 4(–5*), imbricate**, free or partially connate. Stamens 4(–5*), epipetalous or inserted near base of petals and alternate with them; filaments free or connate at base. Anthers 2-thecous; thecae back to back, opening by longitudinal slits. Disk absent or consisting of separate glands between the filaments. Ovary superior, 1–2-locular; style short; ovules 1–2, erect**. Fruit a berry or drupe. Seeds without endosperm; embryo with thick cordate cotyledons.

A small palaeotropical family of three genera, all of which occur in the Flora area, and are often important constituents of the vegetation in arid areas. The fibrillated stems of *Dobera*, and particularly *Salvadora*, are used by Africans of many diverse tribes as toothbrushes.

Small shrubs armed with axillary spines (some herbarium
    sheets may bear pieces devoid of armature, but no diffi-
    culty will be experienced in the field) . . . 1. **Azima**
Unarmed shrubs or trees:
  Petals free; stamens hypogynous, united into a tube at the
    base . . . . 2. **Dobera**
  Petals shortly united at the base; stamens epipetalous . 3. **Salvadora**

1. AZIMA

Lam., Encycl. 1: 343 (1783)

Monetia L'Hérit., Stirp. Nov. 1, t. 1 (1785)
Actegeton Blume, Bijdr.: 1143 (1827)

Glabrous or pubescent dioecious shrubs; stems much branched, sometimes rather scrambling, armed with paired or solitary axillary spines. Flowers small, axillary. Calyx campanulate, shortly 4-toothed or 2-partite. Petals 4, free, lanceolate. Male flowers: stamens 4, free, slightly exserted, alternating with the petals; anthers oblong, nearly equalling the linear filaments; ovary rudimentary. Female flowers: staminodes 4, alternating with the petals; rudimentary anthers sagittate at the base; ovary ovoid, 2-locular; ovules basal, 1–2 per locule but usually solitary; stigma sessile, bilobed, hairy. Fruit a berry, 1–2-seeded.

A small genus of spiny shrubs; about three or four species ranging from South and tropical Africa and Madagascar to India, Malaya and the Philippines.

* According to Baker in F.T.A. 4(1): 21 (1902).
** According to E. Phillips, The Genera of South African Flowering Plants, ed. 2: 573 (1951), valvate aestivation can also occur; also the ovule in *Salvadora* is said to be apical.

D.E.

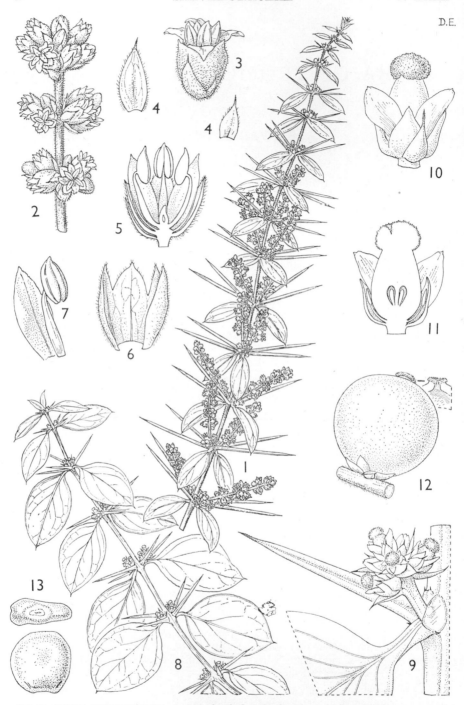

FIG. 1. *AZIMA TETRACANTHA*—**1**, part of male flowering branch, × ⅔; **2**, male inflorescence, × 2; **3**, male flower, × 6; **4**, bracts, × 8; **5**, male flower in longitudinal section, × 6; **6**, calyx, × 8; **7**, petal and stamen, × 8; **8**, part of female flowering branch, × ⅔; **9**, part of node from same, × 3; **10**, female flower, × 8; **11**, same in longitudinal section, × 8; **12**, fruit, showing attachment and separate detail of apex in section, × 4; **13**, seed, two aspects, × 3. 1–7, from *Milne-Redhead & Taylor* 7152; 8–11, from *Faulkner* 1877; 12, from *Verdcourt* 3113; 13, from *Bogdan* 4358.

A. tetracantha *Lam.*, Encycl. 1: 343 (1783) & Illustr., t. 807 (1799); Bak. in F.T.A. 4(1): 22 (1902); Engl., V. E. 3(2): 247, fig. 122/B (1921); F.P.N.A. 1: 511 (1948); T.T.C.L.: 547 (1949); E.P.A.: 486 (1958); Robyns in F.C.B. 9: 234, t. 26 (1960); K.T.S.: 494 (1961) Types.: East Indies, *Sonnerat* & plant cultivated at Paris (P,syn.)

A much branched, tangled, evergreen shrub, 0·6–3 m. tall, with spines up to 3·5 cm. or more long, pale green, glabrous, slightly pubescent or hairy (in some Indian material); branches ± angled, often arching or scandent. Leaves yellow-green; lamina elliptic or oblong to suborbicular, 1·4–5·7(–6·5) cm. long, 0·5–5·6 cm. wide, mucronate (the mucro small but sharp); venation rather prominent. Flowers small, greenish, in small axillary clusters, often running into interrupted spikes at the ends of the twigs. Berries green, turning white, similar to those of mistletoe, ± 0·6–0·8 cm. in diameter. Seeds discoidal, black. Fig. 1.

UGANDA. Bunyoro District: Bugungu, Bulisa, plain bordering Lake Albert, Apr. 1940, *Eggeling* 3864!; Toro District: Mweya peninsula, 19 July 1956, *Brooks* 40! & Katwe, 12 Sept. 1941, *A. S. Thomas* 3957!
KENYA. Masai District: Amboseli, Ol Tukai, 14 May 1961, *Verdcourt* 3113!; Mombasa I., 28 Jan. 1953, *Drummond & Hemsley* 1051!; Kilifi District: Malindi, 25 Dec. 1954, *Verdcourt* 1179!
TANGANYIKA. Tanga District: Mtimbwani, 30 June 1960, *Semsei* 3052!; Pangani District: N. of Pangani, 20 Jan. 1937, *Greenway* 4871!; Rufiji District: Mafia I., Miewi Kubwa, 5 Sept. 1937, *Greenway* 5231!
ZANZIBAR. Zanzibar I., Yozani, 26 Nov. 1930, *Greenway* 2585! & Mbweni, 22 Jan. 1930, *Vaughan* 1126!; Pemba I., Mvumoni, 19 Aug. 1929, *Vaughan* 534!
DISTR. **U**2; **K**5–7; **T**1–3, 5, 6, ?8; **Z**; **P**; widespread from Arabia and the Somali Republic through East and Central Africa to South and South West Africa, Madagascar, Aldabra and Comoro Is., extending to India, Ceylon and Philippines
HAB. In scrub on eroded ground, and particularly on ± saline soils near lakes and seasonal rivers, also in coastal bushland (often not far from high-water mark); 0–1120 m.

SYN. *Monetia barlerioïdes* L'Hérit., Stirp. Nov. 1, t. 1 (1785). Type: a specimen grown at Paris from seeds collected in India by *Sonnerat* (P, holo., ? BM, iso. !)*
    Azima tetracantha Lam. var. *laxior* C. H. Wright in Fl. Cap. 4(1): 490 (1907). Type: South Africa, Natal, East London Park, *M. Wood* 3129 (K, lecto. !)

NOTE. Spineless shoots of this are very easy to mix up in the herbarium with *Salvadora persica* L. If fruits are present, the larger size of those of *Azima* will readily distinguish it.

2. DOBERA

Juss., Gen.: 425 (1789); Verdc. in K.B. 19: 155 (1964)

Tomex Forsk., Fl. Aegypt.-Arab.: CV & 32 (1775), *non* L.
Schizocalyx Hochst. in Flora 27, Beibl.: 1 (1844)
Platymitium Warb. in P.O.A. C: 279, t. 31 (1895); Solereder in Ber. Deutsch. Bot. Ges. 14: 264 (1896)

More or less glabrous, unarmed, evergreen trees or shrubs. Flowers ⚥, sessile in axillary or terminal panicles. Calyx ovoid, irregularly 3–4(–5)-toothed, sometimes approximately bilobed. Petals 4(–5), free, elliptic, oblong or linear-oblong, sometimes ± spathulate, imbricate. Stamens 4(–5), hypogynous; filaments dilated basally, shortly connate into a tube, free and filiform above; anthers ovate. Disk consisting of 4 glands alternating with the stamens, each at the base of a petal and ± attached to it. Ovary 1-locular; style short; stigma obtuse or truncate; ovules 1–2**, erect. Fruit a subglobose or ellipsoidal drupe; only 1 of the ovules appears to develop into a seed in 2-ovuled ovaries.

* A specimen at Kew from the Gay Herbarium was also prepared from a plant cultivated in Paris.
** Three ovules have been found in a specimen from **K**7, Garissa.

A genus of only two species, both occurring in tropical Africa and one extending to India. The two species are closely related but are certainly distinct. In practice they are much more easily separated than the measurements in the following key suggest. Numerous specimens have been dissected. The species overlap in eastern Kenya (see map in K.B. 19: 159 (1964)), but maintain their distinctness.

Ovule solitary; calyx often bilabiate, ± (2·2–)2·8–
3 mm. long; petals (3·5–)4·5–5 mm. long; filaments (1·2–)2–2·5 mm. long; anthers 1·6–2 mm.
long; fruits ± 2 cm. long　.　　.　.　.　.　1. *D. glabra*
Ovules 2; calyx often shortly 4-toothed, ± 2 mm.
long; petals 3–3·5 mm. long; filaments 0·8–1 mm.
long; anthers 1–1·5 mm. long; fruits ± 1–1·5 cm.
long　.　　.　.　　.　.　.　.　.　2. *D. loranthifolia*

1. **D. glabra** (*Forsk.*) *Poir.* in Lam., Dict., Suppl. 2: 493 (1812); R. Br. in Salt, Voy. Abyss., app.: 63 (1814); A. DC., Prodr. 17: 31 (1873); Engl., V.E. 3(2): 249, fig. 123 (1921); Chiov., Fl. Somala 1: 215 (1929); F.P.S. 2: 287, fig. 102 (1952); E.P.A.: 486 (1958); K.T.S.: 496, fig. 90 (1961); Verdc. in K.B. 19: 155 (1964). Type: Yemen, Wadi Surdûd, *Forsskål* (C, holo.)

A much-branched evergreen shrub or tree, up to 1·8–7·5 m. tall; bark green to dark grey, fissured into rectangular patches; wood pale butter-yellow when cut. Leaves olive-green, opposite; lamina coriaceous, elliptic to ovate or obovate, rarely lanceolate or orbicular, 1·5–9 cm. long, 0·7–5·6 cm. wide, mostly obtuse but often acute, usually mucronulate, glabrous or if glands present, then extremely minute but dense; venation often obscure, though sometimes prominent and reticulate; petiole 3·5–4 mm. long. Flowers white, 3·5–5 mm. long (mostly large in the Flora area but smaller in India and Arabia), in axillary and terminal panicles. Calyx minutely papillate. Staminal tube sometimes with small teeth alternating with the upper free parts of the filaments. Ovule solitary in the ovary. Fruit oblong-ellipsoid, (1·8–)1·9–2·4 cm. long, 1–1·3 cm. across. Seeds ovoid, flattened, 12 mm. long, 7·5 mm. wide. Fig. 2/11–14.

UGANDA. Karamoja District: Kidepo valley, Feb. 1960, *J. Wilson* 836!
KENYA. Northern Frontier Province: Moyale, 30 July 1952, *Gillett* 13651 ! & Mandera, 31 May 1952, *Gillett* 13406!; Kitui District: 24 km. E. of Endau, Dec. 1953, *L. C. Edwards* in *E.A.H.* 12315!
DISTR. U1; K1, 2, 4, 7; throughout the dry parts of NE. Africa to the Sudan Republic, Eritrea, Ethiopia and the Somali Republic, extending to S. Arabia and India (Bombay) where it is apparently rare
HAB. Thornbush or scrub of *Acacia*, *Balanites* and *Commiphora*, rocky hillsides and saline river beds; usually on sandy, alluvial, black cotton or calcareous loamy soils; 20–1100 m.

SYN. *Tomex glabra* Forsk., Fl. Aegypt.-Arab.: 32 (1775)
　　Schizocalyx coriaceus Hochst. in Flora 27, Beibl.: 2 (1844); A. Rich., Tent. Fl. Abyss. 1: 108 (1847). Type: N. Ethiopia, Meda valley, *Schimper* 1744 (B, holo. †, K, iso.!)
　　Salvadora glabra (Forsk.) Baill. in Adansonia 9: 290 (1870)
　　Dobera coriacea (Hochst.) A. DC., Prodr. 17: 31 (1873)
　　[*D. macalusoi* sensu Fiori in Chiov., Result. Sc. Miss. Stef.-Paoli 1: 112, t. 24/A (1916), *non* Mattei]

NOTE. *Dobera roxburghii* Planch. (in Ann. Sci. Nat., sér. 3, 10: 191 (1848)), used by J. G. Baker in F.T.A. 4(1): 21 (1902) for this species, is based on some quite different plant which it has not been possible to identify (see K.B. 19: 155 (1964)).

2. **D. loranthifolia** (*Warb.*) *Harms* in E. & P. Pf., Nachtr. zu II–IV: 282 (1897); Bak. in F.T.A. 4(1): 22 (1902); Engl., V.E. 3(2): 249 (1921); T.T.C.L.: 547 (1949); Verdc. in K.B. 19: 157 (1964). Types: Tanganyika,

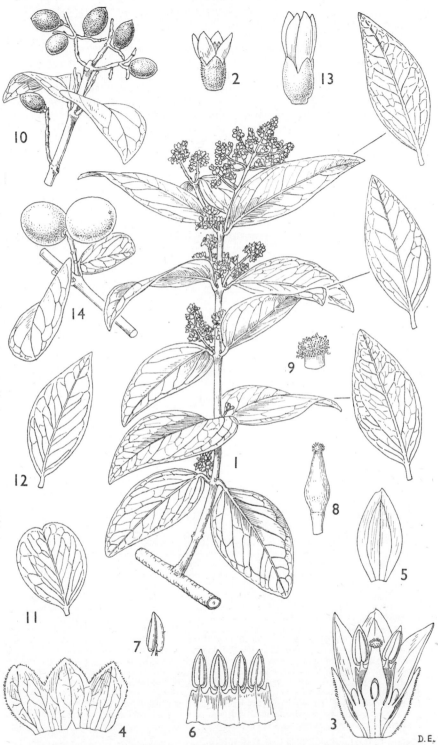

FIG. 2. *DOBERA LORANTHIFOLIA*—**1**, part of flowering branch, × ⅔; **2**, flower, × 4; **3**, longitudinal section of same, × 8; **4**, calyx, opened out, viewed from inside, × 8; **5**, petal, × 8; **6**, stamens, with tube opened out, viewed from inside, × 8; **7**, anther, outer aspect, × 8; **8**, gynoecium, × 8; **9**, style and stigma, × 20; **10**, part of fruiting branchlet, × ⅔. *D. GLABRA*—**11, 12**, leaves, showing variation in shape, × ⅔; **13**, flower, × 4; **14**, part of fruiting branchlet, × ⅔. 1–9, from *Greenway* 10853; 10, from *Bally* 2191; 11, 12, 14, from *Gillett* 4301; 13, from *Gillett* 13651.

W. Usambara Mts., Mashewa, *Holst* 8796* (B, syn. †, BM, K, isosyn. !) &
Kenya/Tanganyika border, Lake Jipe, *Volkens* (B, syn. †, BM, isosyn. !)

A much-branched evergreen tree, 5–15 m. tall, with a bushy crown; bark
variable, black and flaky, pale grey, fissured and rough, or, less often, grey
and smooth.   Leaves opposite; lamina thin or subcoriaceous, elliptic,
obovate or more usually diamond-shaped, rarely almost round, 1·5–9·2(–10)
cm. long, 1·2–4·4(–7·3) cm. wide, mostly acute, shining and quite glabrous or
glaucous with a covering of minute glandular papillae on the lower surface;
venation nearly always prominent and reticulate; petiole 4–5 mm. long.
The leaves frequently bear small hard pea-shaped galls which are very charac-
teristic. Flowers small, greenish-white or white, ± 3 mm. long, in axillary
and terminal panicles. Calyx minutely papillate. Staminal tube sometimes
with small teeth alternating with the upper free parts of the filaments.
Ovules paired in the ovary but only 1 develops in the fruit. Fruit ellipsoid,
1–1·4(–1·6) cm. long, 7–10 mm. across. Seeds dark, ovoid, somewhat com-
pressed, 9–11 mm. long, 6–7 mm. across. Fig. 2/1–10, p. 5.

KENYA. Northern Frontier Province: Kolbio, 29 Aug. 1945, *J. Adamson* 121 !;
    Kilifi District: Kibarani, 21 Mar. 1946, *Jeffery* 499 !; Tana River District: Garissa,
    26 Dec. 1942, *Bally* 1989 !
TANGANYIKA. Lushoto District: Mfumba Steppe, Mnazi, 22 Dec. 1929, *Greenway*
    1972 !; Uzaramo District: Kisiju, Sept. 1953, *Semsei* 1370 !; Lindi District: lower
    slopes of Rondo escarpment approaching Lake Lutamba, Nov. 1953, *Eggeling* 6723 !
DISTR. K1, 4, 7; T3, 6, 8; Somali Republic (S.), Mozambique
HAB. Wooded grassland, *Acacia* scrub and coastal dry evergreen forest, usually on
    black cotton, sandy or alluvial soils; 0–810 m.

SYN. *Platymitium loranthifolium* Warb. in P.O.A. C: 279, t. 31 (1895); Solereder in
    Ber. Deutsch. Bot. Gesell. 14: 264 (1896)
    [*Dobera roxburghii* sensu Bak. in F.T.A. 4(1): 21 (1902), pro parte, quoad spec.
    Wakefield, non Planch.]
    D. macalusoi Mattei in Boll. Ort. Bot. Palermo 7: 184 (1908); Chiov., Fl.
    Somala 1: 215 (1929) & 2: 283, fig. 163 (1932). Types: Somali Republic (S.),
    Brava, Giumbo and Mogadishu, *Macaluso* 25, 38 & 60 (? PAL, syn.)
    D. glabra (Forsk.) Poir. var. *macalusoi* (Mattei) Fiori in Bull. Soc. Bot. Ital. 1913:
    48 (1913); E.P.A.: 486 (1958)
    D. allenii N. E. Br. in K.B. 1914: 80 (1914) & in Hook., Ic. Pl., t. 3017 (1915).
    Type: Mozambique, Antari (not traced as a locality, and quite possibly a
    vernacular name), *Allen* 95 (K, holo. !)
    D. glabra (Forsk.) Poir. var. *subcoriacea* Engl. & Gilg [place of publication not
    traced]; T.T.C.L.: 547 (1949). Type: Tanganyika, Kilosa, *Brosig* (B, holo. †),
    e descr. et loc.
    [*D. glabra* sensu Brenan, T.T.C.L.: 547 (1949), pro parte, quoad spec. *Burtt*
    5308, *non* (Forsk.) Poir.]

VARIATION. In the northernmost parts of its range, i.e. in Kenya near the Somali
    border, atypical specimens are found with smooth bark and shorter very glaucous
    leaves, which are densely minutely glandular on their lower surfaces; the key charac-
    ters still hold good even though along the Tana River both species undoubtedly occur.
    Further detailed collecting may well show that in this area the species have different
    ecological requirements. Judging by Chiovenda's useful comparison (Fl. Somala 1:
    216 (1929)), and assuming he saw the types, *D. macalusoi* Mattei is certainly a
    synonym of *D. loranthifolia*; his mention of very small anthers 0·8 mm. long and
    glandular papillae shows that there can be no doubt. This northern form with glandular
    leaves could possibly be separated off as a variety using Mattei's name. Similar glan-
    dular forms occur in Tanganyika (Tanga), but the amount of this indumentum varies;
    much material from Tanganyika has entirely glabrous leaves with a more prominent
    venation. *D. allenii* N. E. Br. was said to differ by having broader leaves, obtuse petals,
    a larger staminal tube (though analysis says subduplo breviore) and teeth between the
    filaments. All these characters vary; teeth may be present in both *glabra* and
    loranthifolia. N. E. Brown's plant has prominently veined, entirely glabrous leaves and
    the name would be available if this form were ever considered worthy of varietal
    status.

    * Since both the sheets at K and BM, having every appearance of being isotypes, bear
this number 8796 it is assumed that the original citation by Warburg as 8496 is an error.

3. **SALVADORA**

L., Sp. Pl.: 122 (1753) & Gen. Pl., ed. 5: 58 (1754)

Glabrous or pubescent unarmed trees or shrubs, often scrambling. Flowers ♂*, small, numerous, sessile or pedicellate in axillary or terminal panicles. Calyx campanulate, 4-toothed. Corolla campanulate; lobes imbricate, elliptic, obtuse, shortly joined at the base. Stamens 4, inserted at the base or middle of the corolla-tube. Disk consisting of 4 glands alternating with the stamens or absent. Ovary 1-locular; style shortly columnar or almost obsolete; stigma subpeltate or broadly truncate; ovule erect, solitary. Fruit a globose drupe.

A small genus of about four species ranging from Africa and Arabia to India and China. A single species in East Africa whose fruits are much sought after by monkeys and birds.

S. persica *L.*, Sp. Pl.: 122 (1753); Bak. in F.T.A. 4(1): 23 (1902); Engl., V.E. 3(2): 250, fig. 122/A, G–J (1921); Hutch., Fam. Fl. Pl. 1: 237, fig. 180 (1926) & ed. 2, 1: 313, fig. 171 (1959); Sleumer in E. & P. Pf., ed. 2, 20B: 238, fig. 74/A, G–J (1942); T.T.C.L.: 547 (1949); F.P.S. 2: 287, fig. 103 (1952); I.T.U., ed. 2: 371 (1952); E.P.A.: 487 (1958); F.W.T.A., ed. 2, 1: 644 (1958); K.T.S.: 496 (1961); F.F.N.R.: 221 (1962); Verdc. in K.B. 19: 147 (1964). Types: Persian Gulf, *Garcin* (ubi?); India, *Linnean Herbarium* No. 164.1 (LINN, lecto.!)

An evergreen shrub with grey or whitish stems forming tangled thickets, or a small tree, up to 2·7–6 m. tall, glabrous or pubescent. Branches often pendulous, semiscandent, the flowering ones frequently hanging vertically for up to 1 m. Leaves subsucculent; lamina coriaceous, lanceolate to elliptic, sometimes orbicular, 1·4–10·5 cm. long, 1·2–3(–7·5) cm. wide, rounded to subacute or acute at apex, mucronate, cuneate to subcordate at base; petiole 0·3–1·3(–2) cm. long. Flowers small, greenish-white, in numerous lateral and terminal panicles, up to 10 cm. long, with slender racemose branches. Drupes red or dark purple when ripe. Fig. 3, p. 8.

KEY TO INTRASPECIFIC VARIANTS

Plant glabrous:
  Leaves slightly succulent, usually narrower with
    pointed or less rounded apex; usually growing
    in dry inland habitats . . . . .   var. **persica**
  Leaves succulent, broadly elliptic with very rounded
    apex; plant of littoral communities . .   var. **cyclophylla**
Plant pubescent . . . .   var. **pubescens**

var. **persica**

Plant glabrous; leaves slightly succulent, very variable but not narrowly lanceolate nor almost round and very blunt.

UGANDA. Karamoja District: Matheniko, Sept. 1943, *Dale* 355!
KENYA. Turkana District: Lake Rudolf, Ferguson's Gulf, 17 May 1953, *Padwa* 161!; Masai District: Magadi, Aug. 1958, *D. B. Thomas* 726!; Kilifi District: Kibarani, 25 Feb. 1946, *Jeffery* 476!
TANGANYIKA. Shinyanga District: Usanda, Nov. 1938, *Koritschoner* 1741!; Pare District: Mkomazi, May 1939, *Gillman* 750!; Dodoma District: Mwitikira, 16 Aug. 1928, *Greenway* 778!
DISTR. U1; K1–7; T1–6, 8; Arabia, Somali Republic, Sudan Republic, Ethiopia and Eritrea, extending to Mozambique, Malawi, Zambia and South West Africa, Angola and West Africa, also in North Africa extending through the Middle East to NW. India and Ceylon

* In several works the flowers are described as "subunisexual".

FIG. 3. *SALVADORA PERSICA* var. *PERSICA*—**1,** flowering branch, × ⅔; **2,** detail from lower surface of
leaf, × 8; **3,** flower (with pedicel and small segment of rhachis), × 6; **4,** calyx, opened out, × 8; **5,**
corolla and androecium, opened out, × 8; **6,** gynoecium and pedicel, side view, × 8; **7,** same, viewed
from above, × 8; **8,** fruiting branch, × ⅔; **9,** fruit, × 6. All from *Verdcourt* 3581.

HAB.   On saline, sandy or loamy soils, along large seasonal rivers in thorn scrub and
thickets and on desert flood plains, usually associated with *Commiphora* spp. and
Acacia mellifera; also in grassy savannah on alkaline "mbuga" margins, and at the
coast; 0–1350 m.

NOTE.   In Arabia it sometimes forms clumps up to 30 m. in diameter.

var. **cyclophylla** (*Chiov.*) *Cuf.*, E.P.A.: 487 (1958). Type: Somali Republic (S.), Bur
Gao, *Senni* 145 (FI, holo. !)

Plant glabrous; leaves more succulent, broadly elliptic to oblong-elliptic, broadly
rounded at apex, 4–7 cm. long, 2·8–4·5 cm. wide.

KENYA.   Kwale District: Gazi area, unknown collector in *C.M.* 242 ! & 19 km. S. of
Mombasa, Twiga, 1 Apr. 1963, *Verdcourt* 3599 !; Kilifi District: Kikambala, June
1962, *Birch* 62/170B; Lamu District: Kiungamini I., 27 July 1961, *Gillespie* 65 !
DISTR.   **K7**; Somali Republic (S.)
HAB.   On coral just above high-tide level

SYN.   *Salvadora cyclophylla* Chiov., Fl. Somala 2: 282, fig. 161, 162 (1932)

VARIATION.   Although a distinctive variant the leaf-shape is not completely correlated
with the littoral habitat. Plants with narrow pointed leaves also occur at the coast.

var. **pubescens** *Brenan* in K.B. 4: 90 (1949); T.T.C.L.: 548 (1949). Type: Tanganyika, Dodoma District, Kisigo R., *Greenway* 802 (K, holo. !, BM, EA, iso. !)

Branches, inflorescence-axes and leaves softly pubescent; inflorescence open;
leaves slightly succulent, very variable but not round nor very blunt.

KENYA.   Masai District: 82 km. Nairobi–Magadi, 10 Mar. 1951, *Greenway* 8507 ! & Ol
Lorgosailic plains, 2 Aug. 1943, *Bally* 2640 !
TANGANYIKA.   Lushoto District: Mombo, Oct. 1905, *Zimmermann* 959 !; Mpanda
District: Kipangati [Kipengate], 29 Aug. 1959, *Richards* 11391 ! (not so hairy as type);
Mpwapwa District: Kimagai Lake, 30 Aug. 1930, *Greenway* 2487 ! & 2488 !
DISTR.   **K6**; **T2–5**; Zambia, Rhodesia and Angola; less pubescent forms have been
seen from Eritrea and Arabia and are frequent in Central Africa
HAB.   Banks of seasonal rivers, on fine sand, silt or loam soils; 480–1020 m.

VARIATION.   Brenan's remark that this variety is clearly definable and that the species
is otherwise glabrous throughout its range is not true; I have noted tendencies to
pubescence throughout tropical Africa and also in Arabia. Nevertheless, specimens
from around the type area are more hairy than any others I have seen. Some specimens
from **K1** and **K4** have slightly pubescent leaves and stems but I have not cited these.
Other variants occur in Kenya. *Bally* 5639 (**K1**, Mt. Kulal) is close to var. *angustifolia*
Verdc. (not to be confused with *S. angustifolia* Turrill, a fairly distinctive species
occurring in Madagascar, the Aldabra Is. and the Cosmoledo Is.). Much more
material of these variants is required. The variation found in the species is discussed
by Verdcourt (K.B. 19: 147 (1964)).

INDEX TO SALVADORACEAE